GW01161706

# Clever Trevor knows his numbers

## ALAN ROGERS

**GRAFTON BOOKS**
A Division of the Collins Publishing Group

LONDON GLASGOW
TORONTO SYDNEY AUCKLAND

Clever Trevor is a ...

boy

one boy (how many girls?)

Clever Trevor has a ...

sock

two socks

Clever Trevor has a …

guitar

3

three guitars

Clever Trevor has a ...

fish

4

four fishes

Clever Trevor has a …

boat

five boats

Clever Trevor knocks on a ...

door

six doors

Clever Trevor has a ...

mug

seven mugs

Clever Trevor knows a ...

# penguin

8

eight penguins

Clever Trevor has an ...

BONK

apple

9

nine apples

Clever Trevor holds up a …

# finger

10

ten fingers

1

2

*now see if you know your numbers.....*

5    1   2   3   4   5

6    1   2   3   4   5   6

9

3

4

7

8

10

bye

Grafton Books
A Division of the Collins Publishing Group
8 Grafton Street, London W1X 3LA

Published by Grafton Books 1986
Copyright © Alan Rogers 1986

*British Library Cataloguing in Publication Data*

Rogers, Alan, 1952–
Clever Trevor knows his numbers.
1. Numeration—Juvenile literature
I. Title
513'.2   QA141.3

ISBN 0-246-12848-8

Printed in Belgium by
Henri Proost & CIE PVBA

All rights reserved. No part of this publication may be reproduced, stored in a retrieval system, or transmitted, in any form or by any means, electronic, mechanical, photocopying, recording or otherwise, without prior permission of the publishers.